Flanders' Motorcycles

History of the Flanders "4" Motorcycle
1911 - 1914

Larry R. Hierstetter

CMC Press Publications

Flanders' Motorcycles

Copyright © 2023 by Larry R. Hierstetter, All Rights Reserved.

Names:	Hierstetter, Larry R., author.
Title:	Flanders' motorcycles : history of the Flanders "4" motorcycle, 1911-1914 / Larry R. Hierstetter.
Other titles:	Flanders four.
Description:	First edition.	[Westminster, Maryland] : Press Publications, [2023]	Includes bibliographical references.
Identifiers:	ISBN: 979-8-218-14246-9 (paperback)	LCCN: 2023913560
Subjects:	LCSH: Motorcycles--United States--History.	Motorcycling--United States--History.	Flanders, Walter E. (Walter Emmett), 1871-1923--Biography.	Automobile engineers--United States--Biography.	Automobile industry and trade--United States--History. |LCGFT: Biographies.	BISAC: TRANSPORTATION / Motorcycles / History.	HISTORY / United States / 20th Century.	BIOGRAPHY & AUTOBIOGRAPHY / Business.
Classification:	LCC: HE5616.5 .H54 2023	DDC: 629.227/509--dc23

CMC Press Publications

cmcpresspublications@gmail.com

This book was inspired by a brief moment in my grandfather's life as passed down to me by my mother, and written with the goal of better understanding the times in which he lived.

Dedicated to Clyde, Ginny, Shelley, Megan & Matt.

Contents

Introduction	1
1. The Dawn of Motorcycling in the U.S.	5
2. Walter E. Flanders	15
3. Flanders' Automobile Companies	19
4. The Flanders "4" Motorcycle	31
5. A Landmark Motorcycle Factory	49
6. Economic Realities	53
7. Walter Flanders' Untimely Passing	63
8. Special Homecoming	67
Appendix A: Surviving Flanders Motorcycles	75
Appendix B: Motorcycles of the Brass Era	79
Appendix C: Motorcyclepedia Museum	87
References & Resources	93

Introduction

In the spring of 1913 as a young man of 19 years old, my grandfather, Clyde Ecker purchased a used 1912 Henderson motorcycle with a rebuilt engine from a local dealer. The condition of the bike soon became obvious as the machine experienced a catastrophic engine failure during his ride home. Disagreements with the dealer soon led to a lawsuit. Seventy-one pages detailing the proceedings were typed by a stenographer and bound for posterity. As plaintiff, my grandfather's attorney presented seven witnesses, the defendant's counsel presented three. Though an accomplished rider with several years of experience, the opposing attorney attempted to question my grandfather's knowledge and abilities. Clyde responded citing experience with several motorcycle brands, including one that was unknown to me, the Flanders. Rediscovering the court documents more than 107 years later, the mention of a motorcycle brand that my grandfather had ridden, that I had never heard of piqued my interest, inspired a fair amount of research, and inspired this book.

An enthusiast from the earliest days of motorcycling, Clyde continued his pursuits as a young man, eventually purchasing a 1913 Harley-Davidson Model 9E and opening his own motorcycle sales and repair shop in Westminster, Maryland. He went on to marry my grandmother in 1914, establish a loving family of 6 children, and pursue careers as both a state roads inspector and a gentleman farmer during and after the Great Depression.

This book was created to provide a glimpse into the origins of motorcycling in the U.S. The earliest machines were bicycles fitted with small gasoline engines. Technology and engineering progressed rapidly and in a few short years, motorcycle racers traveled at speeds over 100 miles per hour. Methods of manufacturing evolved from small shops turning out hand-built machines to substantial factories mass-producing bikes on increasingly efficient assembly lines.

More specifically, in this book, the focus is on the dynamic and fascinating business and engineering career of Walter Emmett Flanders. Many are familiar with the names of Henry Ford, David Buick, Ransom Olds, and Louis Chevrolet, but the name Walter Flanders is rarely mentioned. A true innovative giant in the early years of the American motoring industry with influential ties to Ford, E-M-F, Studebaker, and Maxwell, Flanders' name is largely unknown. Walter had an immense impact on the young and burgeoning automobile industry, laying the groundwork for some of the most successful companies of the era. In this book, we delve into his business venture attempting to create commercial success with a machine modeled after the "every man's motorcycle" concept of high-quality, reliable, low-cost transportation.

In the pages ahead we'll go back in time to highlight Mr. Flanders' career, his manufacturing genius, and his amazing achievements. More specifically we'll delve into the story of the short-lived Flanders "4" motorcycle brand and where surviving Flanders "4" machines can be seen today.

Chapter 1

The Dawn of Motorcycling in the U.S.

From Rudimentary Designs to Record Breaking Speed within a Decade

The earliest years of motorcycle manufacturing in the United States were crowded with inventors, engineers, financiers, and tinkers, all fighting for financial survival and striving for long-term viability.

The first production motorcycles were built by bicycle manufacturers. Small gasoline engines were fitted into place with rudimentary transmissions, fuel tanks, operating linkages, and ignition systems. Most motorcycles had nothing more than bicycle-derived coaster brakes, none of which provided reasonable stopping power. By 1910, most machines offered sturdier frames, power transmission to the rear wheel by leather belt or chain drive, and some offered magnetos instead of battery-powered ignitions. This was the Brass Era of motorcycle and automobile manufacturing in the United States; so named for the prominent use of brass for fittings, lights, valves, and piping during the time, generally considered to encompass the turn of the century through 1915.

In 1901 Indian Motorcycle Manufacturing Company owned by bicycle manufacturer George M. Hendee began producing motorcycles in Springfield, Massachusetts. 1902 brought production in excess of 500 new machines. In 1913 Indian produced more than 32,000 motorcycles, its highest annual production number ever.

In 1903 William Harley with his partners, brothers Arthur and Walter Davidson, in a 10 x 14 foot backyard shed in Milwaukee, Wisconsin, launched what would become the greatest motorcycle manufacturing success story in the world. With their earliest machines being low-volume experimental designs, 1905 was the first year of production. In 1907, brother William Davidson joined the fledgling company. By 1920 Harley-Davidson had become the largest motorcycle manufacturer in the United States with dealers in more than 65 countries.

With an original focus on economical transportation, motorcycle riding soon grew into a hobby for many. Motorcyclists quickly joined together for recreation, organized rides, and Gypsy Tours offering prizes for achieving specific goals during pre-planned routes, and of course, for racing.

As motorcycling became a more popular hobby, racing quickly turned into an extremely competitive sport. Innovation occurred rapidly as manufacturers and consumers hastily adopted technologies created to turn out faster, more robust machines. As motorcycle racing's popularity continued to grow, the need and opportunity for stationary venues that could draw and hold large crowds of spectators became clear to a few enterprising enthusiasts.

Contests of speed and durability moved from city streets, corn fields, and horse racing venues to purpose-built wooden tracks known as motordromes. Compared to the muddy and dusty dirt roads of the time, wooden tracks provided a relatively smooth surface that allowed for a tremendous increase in speed. The breakneck competition that followed became the deadliest form of racing in the history of United States motorsports.

During this time, milled wood was plentiful and cheap. Track designs included 2x4 lumber turned on edge for the race track surface with grandstands constructed above the track. Viewing platforms were often built around the entire raceway. Track angles in the turns ranged from 15 to 50-degree banks, allowing ever-increasing speed and danger. Riding surfaces were treacherous as motorcycles of the day used "total loss" lubrication systems. Riders had to manually pump oil from a holding tank or manually adjust a drip system to provide proper lubrication to the engines' exposed valves and springs. The bikes, by design, almost constantly dripped oil onto the boards.

Several motorcycle brands were familiar on the tracks with Cyclone, Indian, Excelsior, Thor, and Harley-Davidson having factory teams riding custom-built racing machines. Harley-Davidson was intentionally late to the scene, wanting initially to distance the brand from the apparent danger of the sport. Once committed they soon achieved notoriety and great success with their "Wrecking Crew" racing team.

Board track racing bikes were single-gear, clutchless machines requiring that riders and their bikes be push-started by members of their pit crews. With carburetors running wide open, often reaching speeds over 100 mph, the motorcycles had no brakes.

Daredevil racing legends such as Albert "Shrimp" Burns, Eddie Hasha, Ray Weishaar, Morty Graves, Carl Goudy, Ray Creviston, Jim Davis, Maldwyn Jones, and Otto Walker were popular stars of their era. Fully understanding the dangers of their sport, successful riders were paid up to an equivalent of $500,000 per year in today's dollars.

Riding gear of the day offered minimal upper body protection by leather helmets, goggles, gloves, and heavy woolen sweaters. Lower body protection included leather pants, gaiters, and boots.

As popularity of the lightning-fast sport grew, so did the number of racing venues.

The first motordromes were built by Jack Prince & Fred Moscovics. Prince was a bicycle enthusiast and racer. Moscovics was a racing enthusiast, bicyclist, and mechanical engineer. Following is a partial list of board tracks that existed at more than 24 locations in the United States. City, track distance, and years of operation are shown below:

- Los Angeles (Playa Del Rey), California 1.0 mile 1910-1913

- Elmhurst, California 0.5 mile 1911-1913

- Chicago (Maywood), Illinois 2.0 miles 1915-1917

- Tacoma, Washington 2.0 miles 1915-1921

- Omaha, Nebraska 1.25 miles 1915-1917

- Des Moines, Iowa 1.0 mile 1915-1917

- Brooklyn, New York (Sheepshead Bay) 2.0 miles 1915-1919

- Cincinnati, Ohio 2.0 miles 1916-1919

- Uniontown, Pennsylvania 1.125 miles 1916-1922

The second wave of motordromes was built by Jack Prince & Art Pillsbury, an MIT-trained Civil Engineer.

- Beverly Hills, California 1.25 miles 1920-1924

- Fresno, California 1.0 mile 1920-1927

- Cotati, California 1.25 miles 1921-1922

- San Carlos, California 1.25 miles 1921-1922

- Kansas City, Missouri 1.25 miles 1922-1924

- Altoona / Tipton, Pennsylvania 1.25 miles 1923-1931

- Charlotte (Pineville), North Carolina 1.25 miles 1924-1927

- Culver City, California 1.25 mile 1924-1927

- Salem, New Hampshire (Rockingham) 1.25 miles 1925-1927

- Laurel, Maryland 1.125 miles 1925-1926

- Miami, Florida (Fulton-by-the-Sea) 1.25 miles 1926-1927

- Akron-Cleveland, Ohio 0.5 mile 1926-1930

- Bridgeville, Pennsylvania 0.5 mile 1927-1930

- Hammonton, New Jersey (Atlantic City) 1.5 miles 1926-1928

- Woodbridge, New Jersey 0.5 mile 1929-1931

Often drawing crowds of more than 10,000 spectators, board track racing events were a wild but short-lived commercial success. The dangers of ever-increasing speed too frequently exceeded the tracks' designs. Riders, as well as spectators, were at significant risk during every race. Barely a week went by without news of rider or spectator deaths at the tracks. In particular, two horrific crashes led to the beginning of the end of board track motorcycle racing.

On September 8th, 1912 at the motordrome in Newark, New Jersey, after the main racing events were over, popular and record-breaking rider Eddie Hasha (nicknamed the "Texas Cyclone") and world record holder Ray Seymour competed in a 5-mile handicapped race with four other riders. Attempting to make an adjustment to his motor at 92 mph, Hasha lost control of his machine and was instantly killed when he ran off the track and landed in the crowd. Four spectators were killed with 19 others being injured. The riderless motorcycle then returned to the track causing fatal injuries to fellow racer Johnny

Albright. News of the crash made the front page of the New York Times. Having opened earlier that same year the track was closed, never to reopen.

On July 20, 1913, in Ludlow, Kentucky a racer named Odin Johnson suffered a horrific crash when his machine left the track and hit a light pole. The motorcycle's gas tank burst open and exposed electrical wires ignited the fuel. Flames spread into the crowd of spectators. Eight people lost their lives that day.

As time went by an increasing number of popular racers lost their lives to the sport. Albert "Shrimp" Burns died in a crash in Toledo in 1921, Ray Weishaar died during a race in Los Angeles in 1924 and Eddie Brinck died during a race in Springfield, Massachusetts in 1927.

Over the next few years, motordrome racing all but disappeared. Racers returned to dirt tracks as the thrilling but deadly board track venues, by then nicknamed "murderdromes" fell into disrepair, burned, or were dismantled.

Though the board tracks were gone, America's love of motorcycling was by then firmly entrenched in our culture. The groundwork laid by racers and early industrial pioneers had ignited the motoring public's passion.

One such pioneer seemed particularly well set for success....

1912 Flanders Racer

Albert "Shrimp" Burns

Ray Weishaar

Chapter 2

Walter E. Flanders

Manufacturing Genius & Titan of Industry

Walter Emmett Flanders was born on March 04, 1871, in the town of Rutland, Vermont. He was the oldest of three children born to Dr. George T. Flanders (b: January 4, 1845) and Mary M. (Goodwin) Flanders (b: September 10, 1849). The elder Mr. Flanders began his education working alongside a medical doctor in Chelsea, Vermont. He then attended Homeopathic College in Philadelphia, graduating from that institution in 1870. Dr. Flanders moved to Rutland in 1877 to establish his own medical practice. Known to be a friendly, generous-hearted man with a prosperous and growing medical practice, Dr. Flanders passed away at just 38 years of age. At the time of Dr. Flanders' death on October 15, 1883, his son Walter was just 12 years old.

Walter left school at the age of 15 to become a mechanic and machinist. He began his career servicing sewing machines during an apprenticeship with Singer Manufacturing Company in Elizabeth, New Jersey. In the late 1890s, Walter moved to Cleveland, Ohio to work in general machining with Thomas S. Walburn. Flanders became an inventor of machine tools, an expert in their use, and a knowledgeable salesman. Walter knew how to create machinery and how to arrange it for optimal production. Working out of Cleveland, Walter represented three different machinery manufacturers: Potter & Johnson, Maxwell & Moore, and the Landis Tool Company. When machinery was purchased from Flanders, he would travel and be on-site to manage set-up, demonstrate how the equipment worked, and train the men who would become operators. In 1905 during a sales trip to Detroit, Michigan, Flanders stopped by Henry Ford's Piquette Avenue automobile plant. It so happened that Ford needed one thousand crankshafts. Walter accepted Henry's challenge to manufacture them within a tight timeframe. Flanders' successful delivery of a high-quality product in the specified time impressed Henry greatly. As Ford purchased more equipment, Henry continued to observe Flanders setting up machinery in the now-famous plant. Walter rearranged the location of machinery in the Ford factory for logical order of operations and increased efficiency. Impressed with his work, in 1906 Henry Ford hired Flanders as his factory "works manager" with an agreement that Walter could continue to sell machinery outside of the Ford organization. Ford promised a $20,000 bonus if Flanders could deliver 10,000 cars in one year. Two days before the one-year deadline, the 10,000th car rolled off of the assembly line. In his short 20-month time at Ford, Flanders contributed immeasurably by laying the groundwork necessary for the coming era of mass production. Much of the development of the moving assembly line, for which Henry Ford and his Model T became famous, is credited to Flanders.

A large man of perhaps 275 pounds with a booming voice, Flanders worked hard and played hard. Though known as a boisterous hell-raiser off the clock, Walter was punctual every day. With a reputation for being forthright and fair, his men admired him tremendously. In time, Henry Ford became concerned that the powerful force that was Walter Flanders' personality might overtake Ford himself, even inside his own company.

Flanders understood that the future of the automobile industry would be made up of a few organizations of great size rather than the large number of small companies that existed at the time. Walter wanted to control his own future and he knew that dream was not possible working as Ford's production manager. Flanders submitted his resignation in March of 1908. Thomas Walburn, who came to Ford from Cleveland with Flanders, and Max Wollering, a talented machinist whom Flanders brought to Ford Motor Company, both left with him.

With plans already in place for his post-Ford career, Flanders was immediately announced as general manager of the Wayne Automobile Company. Wayne's spokesmen announced that their strategy included automobile production on a scale never before attempted. Flanders was taking his Ford experience to a company where he would wield more control. Just one week later, Ford announced their new Model T. Ford and Flanders were planning to go head to head.

Walter Emmett Flanders
March 4, 1871 - June 18, 1923

Chapter 3

Flanders' Automobile Companies

A FAIR AMOUNT OF planning must have occurred before Flanders' abrupt departure from Ford. Just a short time later it was announced that Walter had teamed up with industry veterans Byron F. "Barney" Everitt and William E. Metzger. With the partners being referred to as "the big three" within the auto manufacturing press, the new firm arrived with instant credibility and industry respect.

Barney Everitt began his career as a carriage maker in Detroit in 1891 at the age of 19. With an excellent reputation in the industry, in 1899 Everitt started his own company making, painting, and trimming automobile body components. Orders from Ransom Olds and Henry Ford helped to further establish his reputation for quality craftsmanship. Around 1904, seeing others' success in the growing industry, Everitt launched his own automobile manufacturing concern, The Wayne Automobile Company (Detroit being in Wayne County, Michigan). Everitt was, at the time, one of the most well-known names in the industry.

William Metzger was a young bicycle merchant whose automotive passion was ignited when he visited the world's first auto show in London, England. Returning to Detroit, he purchased several electric cars, then marketed and sold them. Finding success, William purchased several steam-powered automobiles and sold those. Metzger subsequently established what was likely the first automobile dealership in the United States. William then helped stage America's earliest automobile shows in New York and Detroit. In 1902 he associated himself with the Northern Motor Car Company. Also in 1902, Metzger became one of the founders of Cadillac. A charismatic salesman, he took orders for more than 2,700 autos at the New York show when only three cars had been produced and before the car had a name.

Combining the initials of the founders' last names the E-M-F Company was born. With the new organization essentially being a merger of Everitt's Wayne Automobile Company and Metzger's Northern Motor Car, E-M-F immediately had three ready-made manufacturing plants. The intention of the new company was the manufacture of high-volume, low-priced four-cylinder cars.

With manufacturing capabilities in place, in what turned out to be a consequential decision, E-M-F began to search for a ready-made dealership network. In July of 1908 Everitt, Metzger, and Flanders contracted with the Studebaker Corporation of South Bend, Indiana. Dating back to 1853, Studebaker was a leading manufacturer of horse-drawn wagons who had previously made attempts to enter the growing automobile business. In 1902 Studebaker marketed an electric car that soon failed. In 1904 they marketed a somewhat successful gas-powered automobile. E-M-F believed that successful sales numbers could be achieved based largely on distribution through Studebaker's existing dealer network. With the business contract in place, E-M-F instantly had 4,000 retail outlets in the United States and abroad. Studebaker would handle all E-M-F export business and one-half of all domestic sales, then estimated at 12,000 units per year. Sales in the United States were separated so that Studebaker handled all business in the South and West, with the rest of the country being handled by Metzger of E-M-F.

In July of 1908, the first E-M-F cars were sent down the production line. Deliveries began in September. By the end of the year, 172 E-M-F autos had been produced but all were recalled due to an inefficient thermo-siphon cooling system. With design changes made, including an effective water pump-based system, production was soon restarted.

Initially, there were two automobile offerings: the larger Model 30 and the smaller Model 20.

The Model 30 was powered by a 30hp, 226.2 cubic inch inline four-cylinder engine. The five-passenger car had a three-speed manual transmission, 34-inch wooden spoke wheels, and a wheelbase of 112 inches. The vehicle had oil-burning side and tail lamps and headlights that ran on acetylene gas. Models included a four-door touring car and a two-door roadster.

The Model 20 was powered by a 20hp, 155 cubic inch inline four-cylinder engine and had a wheelbase of 100 inches. The "20" was directly aimed at the Ford Model T, in an attempt to beat Henry Ford at his own game. The Model 20 however was consistently more expensive than the Model T as Ford was regularly able to reduce its price. In the first year of production, there were just two body styles: a two-person runabout selling for $750 and a four-person touring car selling for $790. More model offerings followed as the Suburban replaced the touring car, and a three-person roadster was offered, as was the first closed car in the price range.

In May of 1909, shortly after initial production began, the E-M-F partners were already at odds over the alliance with Studebaker. Metzger, the outstanding businessman and promoter had not liked the sales arrangement from the beginning and Everitt soon joined him in that opinion. Everitt and Metzger sold their E-M-F interest to Studebaker and moved on to create a new automobile company. Flanders became the chief executive officer of Studebaker making the ties between that company and E-M-F even closer.

Initials not comprising their company name a second time, Everitt and Metzger decided to call their new organization the Metzger Motor Car Company with their vehicle model being labeled the Everitt. The new firm was incorporated in September 1909. With well-respected company leadership and close coverage by the press, the initial run of 2,500 cars was pre-sold before production began.

Meanwhile, at predominately Studebaker-owned E-M-F, with Walter as both president and general manager, Studebaker was pushing to merge the two entities. Though Walter held those titles, Studebaker men were placed in most leading management positions at E-M-F, creating animosity in most of the staff. Additionally, Flanders was not happy with the potential sale as it would dethrone him from fully running his own company. Pursuing that desire, Flanders convinced the Studebakers to help him buy the assets of the defunct DeLuxe Motor Company in Detroit, whose fully equipped factory was immediately ready for new production. With that purchase, Walter took 150 of his men to the new location and founded the Flanders Automobile Company. His new factory would be used to produce the Flanders 20 automobile, a vehicle nearly identical to the E-M-F Model 20. This was Flanders' model designed to compete with Henry Ford's Model T. In 1910, the company planned to produce 25,000 cars, however, only approximately 5,000 were made. All Flanders 20 vehicles were sold through Studebaker dealerships. Back at E-M-F, that company decided to cease production of the "20" so that they could focus on production of the successful "30" and to make room in the dealerships' line-ups for Flanders' new automobile.

In March of 1910 Studebaker, with the help of J.P. Morgan, orchestrated a take-over of all remaining E-M-F assets. The previous purchase of Everitt's and Metzger's shares made the transition a relatively easy one. For the buyout, Flanders was paid one million dollars for his stock and he agreed to stay on as general manager of Studebaker Corporation for a term of three years.

With the proceeds from the sale, in January 1911, Walter created the Flanders Manufacturing Company. This new organization was established to consolidate five relatively new businesses, including Pontiac Motorcycle Company, in which Walter owned a major interest. The purpose of the new company was the creation of a value-positioned motorcycle with state-of-the-art design and features.

Also in 1911, at Walter's new Flanders Automobile Company production was running at peak speed and efficiency but unfortunately, sales fell far short of expectations. In Walter's mind, Studebaker was not holding up their end of the deal in terms of promotions and sales. Flander's reportedly made his dissatisfaction emphatically clear to Studebaker. Walter was likely discovering that Everitt and Metzger had been right about the Studebakers from the start. In its three years of production, 30,707 Flanders 20 automobiles were built.

In May of 1912, Flanders continued to hold his full-time position as general manager with Studebaker. While Walter's other ongoing manufacturing interests (the Flanders Manufacturing Company and The Flanders Automobile Company) were not in direct competition with Studebaker, the situation quickly became awkward. During a Flanders Manufacturing Company Board of Directors meeting, Walter himself was elected president. Of interest, Clement Studebaker was reportedly absent from the board meeting that day. As the news was not well received by Studebaker Corporation, Flanders was effectively demoted while the title of corporate general manager was created for a new leader, James Newton Gunn. Flanders was re-classed to general manager of the Studebaker company's "automobile interest," a specific and considerable difference. Given Walter's obvious dissatisfaction with the unfolding situation, rumors soon circulated of his resignation. Flanders did indeed resign within a very short period. Of note, during Walter's tenure at Studebaker/E-M-F from 1909 to 1912, the brand outsold every other U.S. automobile manufacturer except Ford.

Also in May of 1912, the Flanders Automobile Company launched the new model 50-Six, a vehicle nearly identical to Metzger Motor Car Company's Everitt Six-48. Only a few of these 130-inch wheelbase cars were ever made as business relationship changes were occurring rapidly.

In June of 1912, Flanders rejoined his old partners Everitt & Metzger at Metzger Motor Car Company. Later that same month it was announced that the Metzger Motor Car Co. would be succeeded by a new organization called the Everitt Motor Company. By August 1912 both Studebaker Corporation and Walter Flanders had filed lawsuits against each other regarding violations of his employment contract. Perhaps recognizing the futility of the legal impasse, Studebaker agreed to release Walter from his contract and allow him to manufacture automobiles under whatever name he chose. The very next day a press release announced that the Everitt Motor Company would change its name to the Flanders Motor Company. Now combining products and designs, the new Flanders 50-Six automobile was the old Everitt Six-48 with the addition of electric lighting, electric starting, and other minor refinements. Though a new start was in motion, there was bad news on the horizon. With sales at the new Flanders Motor Company missing goals, Walter's finances were under substantial stress. A new opportunity presented by an old friend would be the next chapter in Flanders' automobile career.

The United States Motor Company

Eventually purchased and led by Flanders who joined the firm in late 1912, the United States Motor Company (USMC) was organized in 1910 by automobile man Benjamin Briscoe. USMC began as the International Motor Company in 1908 as an attempt to create a major force within the industry by combining Maxwell-Briscoe and Buick. That effort had failed so the organization was renamed the United States Motor Company, then focusing on a domestic business model. Briscoe was an auto industry veteran who had brought a car to market in 1908. Briscoe's goal was to combine constituent companies to gain market share, name recognition, and financial success. Improved access to funding was perhaps the primary concern and goal of the USMC as several of the member companies were having great difficulty obtaining vital financial backing. A total of 11 companies made up the organization, each headed by their founder. Those companies included: Maxwell, Briscoe Manufacturing, Stoddard-Dayton, Grabowsky Motor Vehicle Company, Brush Motor Car Company, Columbia Automobile Company, Courier Car Company, Alden Sampson Trucks, Riker, Gray Marine, and Providence Engineering Works. Thomas Motor Company and a few other manufacturers were added after the organization was formed. In an announcement preceding the release of their 52 models to be offered in 1911, USMC boasted of 18 manufacturing plants, 14,000 employees, and the capability of producing 52,000 vehicles per year. Things did not go as well as planned. In 1912 USMC suspended dividend payments on preferred stock. In September of that year, the company went into receivership. Reportedly, a conflict between two of the company's financial backers, who also had financial interests in General Motors, led to the downfall. Benjamin Briscoe retired in 1912 and was replaced by Walter Flanders as manager of the receivership-committee.

In a stunning series of events, on December 31, 1912, Walter Flanders created the Standard Motor Company as a shell organization to be used for the purchase of United States Motor Company's assets. USMC was purchased free of debt, along with those of the Flanders Automobile Company. In January of 1913 assets of the USMC were sold at a public foreclosure sale. All assets were purchased by Flanders who reorganized the company as the Maxwell Motor Company, Inc., preserving the name of USMC's only surviving member and most popular car. The Flanders Motor Company plant in Detroit was converted to make the Maxwell Six and the company was valued at $47 million when this was all over, giving Flanders and his backers a substantial profit.

The Maxwell 25 Roadster, launched in the summer of 1913 as a 1914 model, would outlast the Maxwell Motor Company. Maxwell flourished until 1920 when the faltering economy took its toll. More than 70 percent of the company's 34,169 vehicles produced that year went unsold. After continued financial struggles, Walter P. Chrysler took a controlling interest in Maxwell and in 1925 formed the Chrysler Corporation to replace the Maxwell Motor Company. The Maxwell 25 was relabelled in 1926 as the Chrysler Four and in 1928 the car was revamped and renamed as the first Plymouth automobile.

Runabout Seating Two $750 With Extra Seat (Seating 4) as shown above $40 Extra
Prices F. O. B. Detroit

MAGNETO included—of course
5 LAMPS
Tube Horn and Generator

Chapter 4

The Flanders "4" Motorcycle

In the years between 1900 and 1915 there were more than 120 motorcycle manufacturers in the United States, all vying for commercial success. Even with so many companies competing for economic viability, Walter Flanders felt that there was sufficient room in the burgeoning motorcycle market for him and his automobile industry partners. Walter's manufacturing genius, his obvious knowledge of the motoring industry, his savvy marketing team, and his track record of industrial manufacturing success seemingly positioned his product well ahead of its competition. With Flanders at the helm and his name on the machine, the new motorcycle nameplate appeared to be primed for success.

In October of 1910, with cash proceeds from the E-M-F buyout by Studebaker, Flanders founded the Pontiac Motorcycle Company, capitalized at $600,000. Walter was looking to apply his mass-production know-how and proven levels of efficiency to motorcycle manufacturing. While there was little media coverage of the new company, it was known that plans existed for both two and three-wheeled machines. Walter, being a substantial investor in the new company and on its board of directors, was clearly convinced of the immense potential for motorcycle manufacturing in the fast-growing market.

Flanders brought proven leadership to the company by naming Robert M. Brownson as president and general manager. Brownson was formerly Walter's right-hand man at the E-M-F Company as secretary and treasurer. George W. Sherman and Harold VanDeusen, both veterans in motorcycle sales at the Aurora Automatic Machinery Company, also joined Walter in his new venture.

In addition to the motorcycle manufacturing concern, Flanders simultaneously spearheaded the organization of four separate but critically connected companies. The organizations created to support the Pontiac Motorcycle Company by manufacturing vital components included: Grant & Wood Manufacturing Company, the Pontiac Drop Forge Company, the Pontiac Foundry Company, and Vulcan Gear Works. All operations were set up in the same manufacturing complex.

In January of 1911, with a capital stock of $2,500,000, the Flanders Manufacturing Company was formed via a merger of the five interrelated companies. Officers of the new company included Robert Brownson, president and general manager; A.O. Smith, vice president; James Book, Jr., secretary; and Henry Stanton, treasurer. The board of directors consisted of: Walter E. Flanders, president of E-M-F Company; Dr. James Book; John Shaw, president of First National Bank; William Barbour, president of Detroit Stove Works; Arthur Smith, president of A.O. Smith Company; Clement Studebaker, treasurer of the E-M-F Company and of Studebaker Brothers Manufacturing Company; and Robert Brownson. Charles Splitdorf and Charles Palms were added shortly after the initial board was assembled. By mid-February signs painted on the factory buildings had been changed to Flanders Mfg. Co.

Numerous reasons were cited for the merger. For manufacturing, with all purchases being made by one purchasing department, larger single contracts could be made and larger orders given, resulting in lower prices and less clerical work. For supervision, non-productive labor and similar items could be centralized and reduced; tools, jigs, and fixtures could be standardized and parts could be made interchangeable. Plans continued for both a two and a three-wheeled vehicle, powered by a light twin-cylinder engine. Those intentions were soon changed with the focus turning to a single-cylinder machine designed for high-volume sales at a low price, made possible by simple construction and efficient factory operations. The machine was initially positioned as the "Flanders Bi-Mobile" a two-wheeled automobile. The silly description was quickly changed and the Flanders "4" motorcycle was born.

The first Flanders "4" motorcycle rolled off the production line in 1911. The motorcycle featured a single cylinder 4 horsepower motor with magneto ignition, a notable upgrade over the typical battery-powered ignitions more common in the day. Power was transferred to the rear wheel by leather belt and a hand-operated idler pulley tension control lever. Later, a second control lever and a free engine clutch were added. As another premium feature for the day, front forks featured both compression and rebound springs. Included with purchase was a one-year guarantee on workmanship, at the time, unique to the industry.

The motorcycle was a handsome machine, with the frame and tank painted in gloss black, and the Flanders 4 logo in gold outlined in red placed along both sides of the fuel tank. Chrome features were ample and perfectly finished. The bike's tires were bright white rubber. Tiller-style handlebars were capped with black rubber hand grips. A leather tool bag was mounted atop the fuel tank and a leather saddle with a Flanders 4 imprint completed the stunning machine. With a selling price of $175, at least $25 less than any motorcycle of similar quality, the Flanders "4" was advertised as "Affordable to any man making $3 a day."

Information gathered from several print ads for the 1911 model provides a fairly detailed listing of the machine's specifications:

Engine:

- Single cylinder, air-cooled, 4-stroke.

- 29.49 CID (483cc).

- 4 h.p.

- Flanders carburetors were mentioned in the advance catalog, most bikes had Schebler carbs.

- Atmospheric intake, mechanical exhaust.

    ○ The intake valve was held closed by its spring and opened as internal cylinder pressure fell below that of the atmosphere / outside air. There was no mechanical connection or energy applied to open the intake valve.

    ○ The exhaust valve was held closed by its spring and mechanically driven open by a lobe on the camshaft via tappet and rod.

- Total loss lubrication system.

- Oil tank drip feed with manual adjustment and a sight glass. Oil flow was further adjusted with the handlebar grip - the oiler being connected to the carburetor throttle linkage.

- Featuring "Large valves, large bearings, large flywheels, consistent with the bore and stroke."

  - Bore 3.5".

  - Stroke 3.6".

  - Valves.

    - Intake 1 5/8".

    - Exhaust 1 11/16".

- Three ring pistons.

- All bearings were made of phosphor bronze.

- The cylinder and head were cast in one piece. Bored, reamed, and ground.

- Ignition: self-energizing Splitdorf brand magneto was an upgrade for the period. By comparison, less costly battery-powered ignitions were heavier and required maintenance.

- Wheelbase: 53.5 inches.

- Weight: 160-170 lbs.

- Maximum Speed: 50 mph.

- Fuel Use: 85-87 miles per gallon on sandy roads

and steep mountainous terrains with a 220 lb. rider. 100 miles per gallon on good roads.

- Front Forks: compression and rebound dual spring design.

- Frame: highest grade steel tubing of 14 and 16 gauge.
  - Double upper bars, loop style, with a rearward slope, giving a low saddle position.

- Saddle: Troxel brand with the Flanders 4 logo stamped into the leather. Dual rear coil springs or a single front vertical spring.

- Transmission: 1¾" flat leather belt, idler control.
  - Rider adjusted idler pulley lever on the left side of the tank.
  - The lever was pushed forward to put tension on the belt and transfer engine power to the rear wheel. [1]

- Tires: White rubber clincher tires 28 x 2 ½".

- Brake: Thor or Eclipse coaster brake on the rear wheel.

- Headlight: Acetylene gas powered with a supply tank mounted on the frame under the seat (optional). It is unknown if this was a factory option or an aftermarket system.

- Tank: "Whale Shaped" underslung tank holding both gasoline & oil.

- 1½ gallons of gas.
- 1½ quarts of lubricating oil.

- Tool Bag: Leather bag with stamped Flanders logo mounted on the frame above the fuel tank.

Starting & Riding

- Typical of motorcycles of the era, the Flanders "4" was started with the rider first standing beside the machine with the rear wheel lifted off the ground, supported by the bike's rear stand.
  - Gasoline was allowed to flow toward the carburetor by opening the fuel valve on the left side, underneath the tank.
  - The oil knob adjacent to the fuel valve was opened enough to see oil drip past the site glass.
  - The spring-loaded button in the carburetor intake was pulled out slightly and rotated to set the choke.
- With the above steps complete, the rider would mount the motorcycle.
  - The smooth cylindrical tension lever bar on the left (closest to the fuel & oil tank) was pushed forward into ratchet notches to put tension on the leather drive belt.
  - The hand clutch lever on the left (furthest

away from the fuel & oil tank) was pushed to the front to engage the motor with the rear wheel via the drive belt. (See note 1 below)

- The right handlebar grip was twisted counter-clockwise to open the throttle.

- The compression release lever under the right-hand handlebar grip was pulled to open the exhaust valve, making the engine easier to spin for starting.

- Spark timing was advanced by twisting the left handlebar grip clockwise.

- The rider then peddled briskly to spin the rear wheel via the chain on the right side of the bike.

- With the rear wheel spinning and in the air, tension on the belt and the clutch engaged, pedal power spun the engine for starting.

- When combustion began the compression lever was released.

- Once started and running, the hand clutch was pulled to the rear to disengage the engine from the rear wheel, and the choke was closed.

- The coaster brake could then be applied to stop the spinning rear wheel.

- The rider then pushed the motorcycle forward, off of its rear stand, and secured the stand to the clip on the rear fender.

- When ready to ride, the operator slowly pushed the hand clutch forward so the engine engaged with the rear wheel via the belt. With proper throttle and clutch, the motorcycle began to move forward.

- Traveling speed was adjusted by the throttle as desired.

- When ready to come to a stop, the operator pulled the hand clutch to the rear, removing power from the rear wheel, and the coaster brake was applied.

- When intending to park the motorcycle, tension was released from the belt by releasing the tension lever and pulling it to the rear.

- The motorcycle was then dismounted and the rear stand was released from the retaining clip on the fender. With the rider's foot at the rear of the stand, the motorcycle was pulled up and backward to lift the rear wheel off the ground.

THE FLANDERS "4" MOTORCYCLE

FLANDERS "4" MOTORCYCLE $175
Complete with Free Engine Pully and Magneto ...... Price

This is the wonderful Flanders "4" Motorcycle about which you have heard so much. Our representative, Mr. D. W. Smalley, will be at the Fifth Avenue hotel on Friday and Saturday, April 26 and 27. Here's an opportunity for you to examine the machine itself. Be sure and see Mr. D. W. Smalley and secure a demonstration and have him explain the merits of the machine.

FLANDERS MANUFACTURING CO. PONTIAC, MICH.

THE FLANDERS "4" MOTORCYCLE

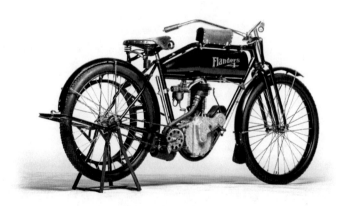

THE FLANDERS "4" MOTORCYCLE

Here is the Flanders "4" Motorcycle

Made in the World's Largest Motorcycle Factory. It is a Big, Powerful, Beautiful Machine, Combining all that is Best and the Price is a little More than Half—$175. Magneto included—of course.

Flanders "4" Motorcycle

$175 f. o. b. Factory---Magneto Included of Course

Has them all on the Run

1. On the earliest 1911 models there was no clutch. Power transmission from the engine to the rear wheel was by belt tensioner pulley adjustment only. These early models can be identified by the lack of the free engine clutch on the outside of the power take-off pulley, the lack of the hand clutch adjustment lever and by the lack of the upper guide that supported the additional lever. The sequence of operations was altered accordingly, providing a less refined riding experience and substantially reduced belt life.

Chapter 5

A Landmark Motorcycle Factory

As Walter searched for a factory to accommodate his motorcycle venture, a prominent, suitable, and vacant facility in Chelsea, Michigan was perhaps the obvious choice. Constructed just a few years prior, what is now called the Chelsea Clock Tower Complex was an ideal location. The building most suitable for motorcycle manufacturing was renamed and occupied by the Pontiac Motorcycle Company.

Built in 1906 the Clock Tower building, alternately spelled Clocktower in local literature, has served many purposes over its lifetime. Today the building remains the prominent architectural feature in Chelsea and often serves as a symbol for the city.

The manufacturing site was developed in 1890 when Frank Porter Glazier and partners created the Glazier-Strong Oil Stove Company to produce stoves with the trademark B&B, for Brightest & Best. Several designs of kerosene and gas stoves were made for both cooking and heating. B&B stoves were sold throughout the United States and around the world.

Glazier eventually bought out his partners and renamed the business the Glazier Stove Company. Two major fires destroyed portions of the growing company, in 1894 and 1895. Damage caused by the fires played a major role in the conceptual design of the reconstructed building. Once the tallest structure in Washtenaw County at 135 feet, the Clock Tower served the dual purposes of local timekeeping icon and of holding a 20,000-gallon water tank for fire suppression. A Seth Thomas clock mechanism with Westminster chimes completed the new structure in September 1907. Four seven-foot illuminated faces were visible from all directions in the city and the tower contained four large bells ranging in size from 300 to 1,250 pounds. The clock mechanism is now operated electrically but for many years it was the duty of the town marshal to climb the stairway to the top of the tower at regular intervals and wind the clock by hand. While the Clock Tower no longer serves as a water tower, the old tank is still encased within the structure just below the clock faces.

In 1907 the Glazier Stove Company became the largest oil stove company in the world, occupying eighteen structures within the ten-acre site. Buildings at the facility included a brass foundry, an iron foundry, a pig iron storage facility, a coal charcoal storage building, a building for enameling and polishing, and several other smaller support structures.

The Clock Tower was the crowning architectural feature of Frank Glazier's company and of his real estate development ventures. The building contained automatic sprinklers, electric elevators, and modern manufacturing machinery. Shortly after construction of the new buildings the Glazier Stove Company went bankrupt amidst a political and financial scandal.

Following the failure of Glazer's company, many other businesses occupied the site and manufactured a variety of products in addition to Flanders' operations there. Lewis Spring & Axle, Chelsea Foundry & Machine Company, Peninsular Furnace Company, Federal Screw Works, Rockwell-Standard Corporation–Spring Division, Central Fiber Products, Chelsea Industries, Inc., and the Real Ale Company have all operated from the iconic structure.

Located at 310 N. Main Street, the 24,410 square-foot building is currently a multi-tenant property containing office space, a mortgage company, attorney's offices, a wedding venue, and several other small businesses. The building was substantially renovated in 2000.

Corner in Finishing Room—Flanders-Chelsea Plant

Chapter 6

Economic Realities

Changes in Ownership & the Last Flanders Motorcycles

In May of 1912, as Walter was running the Flanders Manufacturing Company, he was still employed in his full-time position as general manager with Studebaker. During this tumultuous period, Flanders Manufacturing Company announced that they would build 15,000 motorcycles for the 1913 season. Three models were planned, including an all-new twin-cylinder model and two single cylinder machines, one being the familiar belt-drive model and the other a chain-driven machine, each of the two otherwise identical to the other.

Published plans proved to be extremely optimistic. Sadly the distribution and sales of Flanders "4" motorcycles fell far short of projections. In December of 1912, Flanders Manufacturing Company went into receivership. Walter, being of ample financial means saw the unfortunate sales trends and decided to exit the motorcycle market. Had owning a motorcycle company then been a strong desire, he most certainly would have kept the company afloat.

The straightforward simplicity and economical transportation that automobile drivers of the time desired and found in the Model T automobile, which was the design intent adapted to two wheels for the Flanders "4", turned out to be the opposite of what motorcyclists wanted. While automobile consumers were attracted to utilitarian value, motorcyclists' demands called for power and speed. The concept of the Flanders "4" being an "every man's motorcycle" had largely missed the market. Unexpectedly, it appears that the quality and value proposition of the Ford Model T that Flanders attempted to emulate within the motorcycle industry was one of the larger factors that worked against the bike's success - the coming of even more affordable automobiles. The handsome, sturdy, and well-equipped Flanders "4" motorcycle would live on for a few more brief years, albeit without its namesake and founder.

In mid-1913 the court-appointed receiver handling the Flanders Manufacturing Company's assets approved the sale of the organization to the Harris Brothers Company of Chicago, Illinois. Harris Brothers was a small Chicago-based kit home seller that started as a house demolition company. Originally named Chicago House Wrecking, the name was changed in 1913 to better suit their business operation. Popular in those times was the sale of mail-order house plans. Harris Brothers initially offered a variety of designs, then grew to sell complete house kits. Of the six national companies selling kit homes through mail-order catalogs, Harris Brothers was likely the least well-known.

On October 3rd, 1913, it was published that key employees of the firm were moving from Detroit to the Chelsea plant.

Just two weeks later it was announced that a new company had been organized to continue the manufacture of Flanders "4" motorcycles. Motor Products Company had filed for incorporation with $150,000 in capital stock. The headquarters of the new company would be in Detroit, where administrative functions were already in place. Manufacturing operations would continue at the former Flanders plant in Chelsea. Rather than pursuing a dealer network, mail-order sales were planned with advertising and promotions found only in print publications, at trade shows, via publicity provided by the motoring press, and by word of mouth.

In October 1914, Motor Products Company purchased the Cricket Cyclecar Company of Detroit. All equipment of the Cricket company was shipped to the Chelsea plant for continued production. The Cricket was to be powered by a new Flanders twin-cylinder engine. Cyclecars saw a brief period of popularity in the United States and Europe between 1910 and the early 1920s. They were designed to fill the gap in the market between motorcycles and automobiles. While cyclecars could accommodate two people, passengers were required to sit in tandem, behind the driver. Vehicle bodies were lightweight and the machines offered little weather protection or comfort. With small automobiles continuing to decrease in price, demand for cyclecars fell dramatically. United States brands such as Burrows, Dudley Bug, Kearns LuLu, O-We-Go, and Pacific were soon lost to history. The Cricket was made only in 1914.

For the 1914 model year, in an attempt to reinvigorate the Flanders motorcycle brand a series of upgrades was introduced and the Model B was presented to the public with high expectations. A brochure announcing the new Flanders "4" bikes' release touted:

- Motor - similar to Model A with many improvements.

- New valve construction.
  - New compression release.
  - New crankcase with round corners.
  - Oil gauge.
  - Bore 3½", stroke 3.6".
- Tires: 28" x 2½".
- Transmission: 1¾" flat belt with Eclipse Pulley.
- Handlebars: New design much longer and wider.
- Lamp Bracket on each machine.
- Tank: Larger than Model A and made to fit the frame.
- Ignition: Splitdorf magneto.
- Oiler: Sight feed and grip control.
- Carburetor: Schebler.
- Brakes: New and improved.
- Cranks: New and improved.
- Guards: Extra wide and long.
- Saddle: Troxel.
- Finish: Handsomely nickeled and enameled black.

- Speed: 4 to 50 miles per hour.

- Wheel Base: 53½".

In conjunction with the release of the Model B, a totally new and different design was introduced. Presented as the Model C, the new motorcycle was a radical departure from Flanders' previous design and was unique in the industry.

The 1914 twin-cylinder Model C was chain-driven and featured a large cover concealing a sizable rear sprocket with damper springs. This was a dramatic design change from the handsome lines of previous models. The Model C was described as the only direct chain drive machine with the flexibility of a belt. Fairly large for its day, the 67 cubic inch motor produced 9 hp and was equipped with a Schebler carburetor and Splitdorf magneto, the best components available. Along with the many other differentiating factors from the Model A and Model B, the bright red bike incorporated a completely different and notably less handsome Flanders logo on the tank. Though innovative, the Model C lasted just one year on the market. There are no records showing how many Model C motorcycles were sold. At the time of this writing, there is just one known remaining Model C machine: frame number 7088, engine number 7012.

A Motor Products Company advertisement in Bicycling World and Motorcycle Review Issue #61, 1914 described the Model C (capitalization included):

- Foot band brake - located just left of the floorboard on the right side. Different from other band brakes because of its style, size, and application.

- Foot boards as a part of regular equipment, adjustable to any position.

- Kick starter of our own design, making it possible to spin the motor sufficiently to start immediately under generally favorable conditions. The engine can be started while both wheels remain on the ground.

- Important improvements neatly enclosed on OUR SUCCESSFUL CHAIN and DIRT PROOF HOUSING. THIS feature alone being distinctly and EXCLUSIVE on FLANDERS motorcycles. Its value for the preservation of the chain drive, sprockets, and riders clothes is inestimable.

- DIRECT CHAIN DRIVE and CUSHION SPROCKET and ENCLOSED VALVES of course will be STANDARD and EXCLUSIVE FEATURES on all 1915 models.

- In all, THE MAN WHO KNOWS will recognize in the new Flanders a motorcycle of the HIGHEST CLASS and AT A RIGHT PRICE.

Following are notable excerpts from the Chelsea Standard Newspaper chronicling iterations of the related organizations, 1910 – 1914:

October 1910

Pontiac Motorcycle Company incorporated, capitalized at $600,000. The startling fact is that the company is sponsored by several of the biggest and best known men in the automobile industry. Chief among these is Walter E. Flanders, president and general manager of the E-M-F Company, makers of the E-M-F and Flanders cars. He is a man of large means and a heavy stockholder in the new Pontiac Motorcycle Co. Although the company has just been formed, its sponsors have been engaged for several months in shaping its plan and product. No motorcycle of repute either here or abroad has escaped critical analysis by its experts. It is known that it is the intention of the company to do everything on a large scale and to leave no money unexpended that will push the new motorcycle to the forefront. The factory buildings site also takes in the plants of the Vulcan Gear Works and the Champion Manufacturing Company.

January 1911

Grant & Wood Manufacturing Co. in big industrial merger (including Pontiac Motorcycle Company). With a capital stock of $2,500,000, a new company known as Flanders Manufacturing Company is to absorb the recently organized Pontiac Motorcycle Company, along with four other concerns in which Walter E. Flanders and his associates are the principal stockholders. The concerns include the Grant & Wood Manufacturing, Pontiac Drop Forge Co., Pontiac Foundry, and Vulcan Gear Works. The change involves numerous departures from the tentative program in which the Pontiac Motorcycle concern has been engaged. The planned Pontiac motorcycles will be renamed the Flanders bi-mobile and tri-mobile.

February 16, 1911

The signs of the factory buildings have been changed to Flanders Manufacturing Co.

September 5, 1912

Flanders Manufacturing Company will build 15,000 Motorcycles for the season of 1913. Three models planned. One will be a twin cylinder and the other two will be single cylinder machines, one being chain driven and the other a belt drive. It has been fully demonstrated that the company has the fastest machine on the market today and their finish is second to none. Wednesday afternoon the Chelsea plant was visited by W.E. Flanders.

July 10, 1913

Harris Brothers Company to manufacture motorcycles in Chelsea. On Monday of this week the motorcycle department of the Chelsea plant of the Harris Brothers Company of Chicago, started up and they will give employment to between forty and fifty operators. D.W. Caswell, who was formerly the chief of the engineering department, has been engaged as general manager of the plant and nearly all of the foremen have been engaged and will return here and take their old positions. For the present The Standard is informed that the business will be confined to manufacture of two single cylinder models and the name of Flanders will be retained as the name of the motorcycles. The company will make up a line of motorcycle tools and will also carry on a series of experiments on other lines. The Harris Brothers Company expects to begin shipping out motorcycles about the first of August and the business will proceed forward as fast as possible. The automatic screw machine department will not be placed in operation, but the machinery will (be) sold as soon as possible.

July 24, 1913

The Chelsea Screw Company the last of the past week installed three more automatic screw machines that they purchased from the Harris Brothers. The new company expects to begin operating their plant the last of this week.

July 31, 1913

The Harris Brothers Company made their first shipment of motorcycles on Wednesday of this week. The company has about 40 men at work in the local factory and have nearly 100 motorcycles ready for the market in about a week.

July 31, 1913

Mr. and Mrs. Fred Aichle, who have been residing in Jackson for some time past, have returned to Chelsea. Mr. Auchle has accepted a position in the motorcycle department of the Harris Brothers Company.

October 16, 1913

A $150,000 company organized to manufacture motorcycles in Chelsea. The following appeared in the Detroit Free Press this morning: Organized with a capital stock of $150,000, the Motor Products Company, Chelsea, Michigan, filed articles of incorporation in Lansing yesterday. The company is formed to continue to manufacture the Flanders "4" and "7" motorcycles, which, with a few minor changes and improvements will be turned out exactly the same as the machine formerly built by the Flanders Manufacturing Company. Headquarters of the company will be located in Detroit, where its administrative officers already are transacting business. Operations are in progress at the former Flanders plant in Chelsea, from which shipments are being made.

October 3, 1913

Wm. Miller and T.S. Hughes, who have been working in Detroit for several months, have accepted positions in the Harris Brothers Company motorcycle factory here.

October 8, 1914

The Cricket Cyclecar Company has been absorbed by the Motor Products Company and will be manufactured in Chelsea. The equipment of the Detroit company will be shipped here and used to build the little car. The Cricket is equipped with a Flanders twin cylinder engine. The company will continue to manufacture the Flanders "4" and twin cylinder motorcycles which have been their chief output in the past.

August 26, 1915

The Hudson Motor Co. and the Timkin Axle Co. of Detroit during the past week shipped several pieces of the machinery formerly owned by the Flanders Manufacturing Co. from this place to their plants in Detroit.

Chapter 7

Walter Flanders' Untimely Passing

WALTER EMMETT FLANDERS DIED on June 18, 1923, at just 52 years of age. Walter passed due to injuries sustained in an automobile accident that occurred several days prior. According to friends, he was a passenger in a vehicle that was on its way to his home in Williamsburg, Virginia. Walter sustained a broken leg and several internal injuries, his death was attributed to kidney failure. He was buried at Williamsburg Memorial Park. At the time of his death, it was published that Walter had been married five times and that he had five children.

Walter's former estate of nearly one thousand acres surrounding Green Lake in West Bloomfield, Michigan was later subdivided and developed into lakefront homes and golf courses. In 1949 the Green Lake Association purchased Flanders' elaborate "garage." The Flanders Garage marker placed by the Michigan Historical Commission reads:

Automobile entrepreneur Walter E. Flanders (1871 - 1923) was born on a Vermont farm. At the peak of his success in the 1910's he owned a 1,000-acre estate which included all of Green Lake, large farm outbuildings, a greenhouse, this garage, and adjacent craftsman-style house. Flanders owned a variety of livestock and at times he employed three to four hundred men on his farm. The garage, which used to be for entertaining, has an automobile turntable, a billiard room, a ballroom, and a two-lane bowling alley. Flanders moved to Virginia in 1919. The Aviation Country Club purchased the estate in 1920 and used the garage as a clubhouse. The lakefront land and golf courses have been developed as a subdivision. The Green Lake Association purchased the garage in 1949.

Walter Flanders' obituary from The Chelsea Standard newspaper, Chelsea, Michigan, June 21, 1923:

Walter E. Flanders Dead. Walter E. Flanders, one of the outstanding figures in the Detroit automotive world, and who played a large part in the organization of various automobile concerns in Detroit, died Saturday at his estate in Williamsburg, Va.

Mr. Flanders was superintendent of the first Ford Motor Co. plant in Detroit, going there from Cleveland in 1905.

He was the head of the Flanders Manufacturing Company, which operated the Chelsea plant now owned by the Lewis Spring & Axle Company. The chief output of the plant here under Flanders' management was motorcycles.

Chapter 8

Special Homecoming

An Early Production 1911 Flanders Motorcycle Returns to Chelsea

Fulfilling a long-held dream of many members of the Chelsea Area Historical Society (CAHS) in Chelsea, Michigan, an early production 1911 Flanders "4" motorcycle was returned to the place where it was manufactured, then donated to CAHS in 2017. The motorcycle was located by a CAHS member in the private collection of Virgil Elings in Solvang, California. A small group of enthusiasts collected money for the purchase and returned the bike to Chelsea. The motorcycle was completely disassembled, and all parts were cataloged in detail. Some missing parts had to be fabricated to put the bike back together. The engine, serial number 670, was rebuilt and the motorcycle was returned to running condition on July 16, 2017, the day it was donated to the historical society. Presentation of the Flanders "4" was made possible by many generous local donors and countless hours of work by local experts and enthusiasts. The motorcycle is in beautiful, ridable, original, and un-restored condition.

Special thanks to Dave Strauss of CAHS for providing information on this motorcycle.

Please consider supporting this treasure with membership or donations to:

Chelsea Area Historical Society

128 Jackson Street

Chelsea, MI 48118

734-476-2010

https://www.chelseahistory.org

Appendix A: Surviving Flanders Motorcycles

RESEARCH INDICATES THAT PERHAPS just 12 Flanders "4" motorcycles still exist. While some are held in private collections, below is a list of museums with known survivors on display. Please consider visiting, joining, or donating to these valuable preservationist organizations.

AMA Motorcycle Hall of Fame Museum

13515 Yarmouth Drive

Pickerington, Ohio 43147

(614) 856-2222

info@motorcyclemuseum.org

Barber Vintage Motorsports Museum

6030 Barber Motorsports Parkway

Birmingham, Alabama 35094

(205) 699-7275

bvmm@barbermuseum.org

Chelsea Area Historical Society and Museum

128 Jackson Street

Chelsea, Michigan 48118

(734) 476-2010

membership@chelseahistory.org

Motorcyclepedia Museum

250 Lake Street

Newburgh, New York 12550

(845) 569-9065

coordinator@motorcyclepediamuseum.org

APPENDIX A: SURVIVING FLANDERS MOTORCYCLES

St. Francis Motorcycle Museum

110 East Washington Street

St. Francis, Kansas 67756

(785) 332-2400

Vintage Antique Motorcycle Museum

514 North Market Boulevard

Chehalis, Washington 98532

(360) 748-3472

https://www.nwadventures.us/Motorcycles.html

Note 1: The ELK motorcycle. Wheels Through Time Museum in Maggie Valley, North Carolina has the only known intact and unrestored ELK motorcycle in existence. Oral history says that when the Flanders Manufacturing Company went out of business, a hardware store owner in Elkhart, Indiana bought some of the remaining stock and rebranded them as ELK machines. Careful inspection will prove that the ELK is, without a doubt, a re-badged Flanders "4".

Note 2: Flanders Company, Inc., a current-day motorcycle business located in Duarte, California, owns 2 Flanders motorcycles: a black 1911 and a red 1914. Not relatives of Walter Flanders, this family-owned company is a motorcycle parts and accessories business, not a museum. Special thanks to Mr. Paul Flanders for his information.

Appendix B: Motorcycles of the Brass Era

United States 1900 - 1915

A.M.C. / Allied Motors Corporation (1912–1915) Chicago, Illinois

America (1904–1906) La Porte, Indiana

American (1902–1910) Chicago, Illinois (also sold under the brands Columbia, Tribune, Rambler, and Crescent)

American (1911–1914) Chicago, Illinois

Armac (1902–1913) St. Paul, Minnesota

Apache (1907–1911) Denver, Colorado

Arrow (1909–1916) Chicago, Illinois

Bailey (1913–1917) Portland, Oregon

Bradley (1905–1912) Philadelphia, Pennsylvania

California (1901–1903) San Francisco, California

Champion (1911–1913) St. Louis, Missouri

Chicago (1904–1905) Chicago, Illinois

Clemens (1901–1903) Springfield, Massachusetts

Clement (1903–1909) New York, New York

Cleveland (1902–1905 / 1915–1929) Cleveland, Ohio

Columbia (1902–1905) Chicago, Illinois

Comet (1908–1910) Milwaukee, Wisconsin

Crawford (1913-1914) Saginaw, Michigan

Crescent (1902–1905) Hartford, Connecticut

Crouch (1905–1908) Stoneham, Massachusetts

Curtiss (1901–1913) Hammondsport, New York

C.V.S. / C. V. Stahl Motorworks (1911–1917) Philadelphia, Pennsylvania

Cyclone (1913–1916) St. Paul, Minnesota

Dayton (1911–1917) Dayton, Ohio

Deluxe (1912–1915) Chicago, Illinois

Detriot (1911) Detroit, Michigan

Duck (1903–1906) Stockton, California

Dyke (1903–1906) St. Louis, Missouri

Eagle (1909–1915) Brockton, Massachusetts

APPENDIX B: MOTORCYCLES OF THE BRASS ERA

ELK (1914) Elkhart, Indiana

Emblem (1907–1925) Angola, New York

Erie (1906–1911) Hammondsport, New York

Excelsior Motor Manufacturing and Supply Company (1908–1931) Chicago, Illinois

Feilbach (1912–1913) Milwaukee, Wisconsin

Flanders (1911–1914) Chelsea, Michigan

Fleming (1900–1902) New York, New York

Freyer + Miller (1901–1907) Cleveland, Ohio

Gerhart (1914–1916) Harrisburg, Pennsylvania

Greyhound (1907–1914) Buffalo, New York

Geer (1905–1909) St. Louis, Missouri

Hampden (1901–1903) Springfield, Massachusetts

Harley-Davidson (1903–present) Milwaukee, Wisconsin

Haverford (1909–1914) Philadelphia, Pennsylvania

Hawthorne (1911-1912) Chicago, Illinois

Henderson (1911–1931) Detroit, Michigan

Hercules (1902–1903) Hammondsport, New York

Hilaman (1906–1912) Moorestown, New Jersey

Hoffman (1903-1904) Chicago, Illinois

Holley (1902–1911) Bradford, Pennsylvania

Indian (1901–1953) Springfield, Massachusetts [1]

Iver-Johnson (1907–1916) Fitchburg, Massachusetts

Jefferson-Waverley (1910–1914) Jefferson, Wisconsin

Joerns (1910–1916) St.Paul, Minnesota

Keeting (1901–1902) Middletown, Connecticut

Keifer (1909–1911) Buffalo, New York

Kokomo (1909–1911) Kokomo, Indiana

Lamson (1902–1903) Abington, Massachusetts

Light (1901–1911) Pottstown, Pennsylvania

Mack (1909–1913) Milwaukee, Wisconsin

Maltby (1902–1903) Brooklyn, New York

Manson (1905–1908) Chicago, Illinois

Marsh (1900–1913) Brockton, Massachusetts

Marvel (1910–1913) Hammondsport, New York

Mayo (1905–1908) Pottstown, Pennsylvania

Merkel (1902–1915) Milwaukee, Wisconsin

Miami (1915–1916) Middletown, Ohio

Michaelson (1908–1915) Minneapolis, Minnesota

Militor / Militaire (1911–1922) Cleveland, Ohio; Buffalo, New York; Bridgeport, Massachusetts

Minneapolis (1908–1914) Minneapolis, Minnesota

Mitchel (1901–1906) Racine Junction, Wisconsin

Monarch (1912-1915) Owego, New York

APPENDIX B: MOTORCYCLES OF THE BRASS ERA

Morgan (1901–1902) Brooklyn, New York

Nelk (1905–1912) Palo Alto, California

New-Era (1909–1913) Dayton, Ohio

Orient (1900–1905) Waltham, Massachusetts

Peerless (1912–1916) Boston, Massachusetts

P.E.M. / Waverly Manufacturing Company (1905–1912) Jefferson, Wisconsin

Pierce (1909–1913) Buffalo, New York

Pioneer (1908–1910) Worcester, Massachusetts

Pirate (1912–1915) Milwaukee, Wisconsin

Pope (1911–1918) Hartford, Connecticut; Westfield, Massachusetts

Pratt (1911–1912) Elkhart, Indiana

Racycle (1905–1911) Middletown, Ohio

Rambler (1903–1914) Hartford, Connecticut

Reading-Standard (1903–1922) Reading, Pennsylvania

Reliance (1908–1915) Owego, New York

Royal (1909–1910) Worcester, Massachusetts

Schickel (1912–1924) Stamford, Connecticut

Sears (1910–1916) Chicago, Illinois (built for Sears Roebuck & Co. by Aurora Automatic Machine Company and Spacke)

Shaw (1903–1917) Galesburg, Kansas

Smith Motor Wheel (1914–1924) Milwaukee, Wisconsin

Stahl (1902–1907) Philadelphia, Pennsylvania

Steffey (1900–1905) Philadelphia, Pennsylvania

Thiem (1900–1913) St. Paul, Minnesota

Thomas Auto-Bi (1900–1912) Buffalo, New York

Thor (1902–1917) Aurora, Illinois

Thoroughbred (1904–1905) Reading, Pennsylvania

Tiger (1906–1909) New York, New York

Tiger Autobike (1915-1916) Chicago, Illinois

Torpedo (1907–1910) Whiting, Indiana

Tourist (1905–1907) Newark, New Jersey

Tribune (1903–1914) Hartford, Connecticut

Triumph (1908–1912) Chicago, Illinois; Detroit, Michigan

Wagner (1901–1914) St. Paul, Minnesota

Westover (1912–1913) Denver, Colorado

Whipple (1903–1905) Los Angeles, California

Williams (1912–1916) New York, New York

Williamson (1902–1903) Philadelphia, Pennsylvania

Yale (1902–1915) Toledo, Ohio

APPENDIX B: MOTORCYCLES OF THE BRASS ERA

---

1. Later iterations of Indian are acknowledged but not considered linear from the original: (1963-1977) Los Angeles, California, (1999-2003) Gilroy, California, (2006-2011) Kings Mountain, North Carolina, (2011-present) Medina, Minnesota.

Appendix C: Motorcyclepedia Museum

Conducting research for this book, I visited the Motorcyclepedia Museum in Newburgh, New York. Finding one Flanders "4" motorcycle is a rare occurrence. When I learned that Motorcyclepedia had two in their collection, I began to plan my trip. With more than 600 motorcycles on display, there is likely something interesting there for everyone. For music fans there is Prince's Purple Rain motorcycle and Jerry Garcia's 1984 Harley-Davidson FXST Softail. Movie and TV fans will find bikes that appeared in Tron, RoboCop and other productions, along with Fonzie's Triumph from Happy Days. Also on display is a 1964 Harley-Davidson FL police motorcycle that was in President John F. Kennedy's Dallas motorcade on November 22, 1963. This bike was used in Kevin Costner's production of "JFK". All sorts of other bikes are on display, including the largest collection of Indian motorcycles in the world. The collection and display of antique Harley-Davidson motorcycles is breathtaking. All exhibits are extremely well done.

I was there to see the museum's two Flanders "4" motorcycles. The display did not disappoint. Research would not have been complete without the ability to closely examine these bikes. Exhibits are not to be touched, but for the ability to inspect the bikes closely, follow cable routes, understand linkage design, and view both machines from different angles, I am most grateful.

1913

APPENDIX C: MOTORCYCLEPEDIA MUSEUM

1913

1912

Oil sight glass and Schebler carburetor

APPENDIX C: MOTORCYCLEPEDIA MUSEUM

1913 model on the pedestal, 1912 in the foreground

References & Resources

This References & Resources section is provided for reader use and benefit. Where information and material were ascertained from the listed sources while assembling this work, all usage is considered within the limits of fair use under U.S. copyright law (Section 107 of the U.S. Copyright Act). This knowledge has been assembled primarily for scholarship and research. The nature of this work is factual, not creative. Further, this work has been evaluated via the fair use assessment checklist made available by the Copyright Advisory Office of the Columbia University Libraries, 535 West 114th St. New York, NY 10027. A copy of the completed checklist is available from the author.

Andrews, Jill. Chelsea Update. https://ChelseaUpdate.com (2019)

https://architecturalobserver.com/the-chicago-house-wrecking-companys-house-design-no-160/

The Bicycling World and Motorcycle Review Issue #61. (1914)

Brown, Jac. Michigan Sport Touring Riders newsletter. (2017)

Brown, Jac. Lehman Hill Blogspot. My Town's Motorcycle History – The Flanders Motorcycle Company. (2016)

Burgess Wise, David. *Historic Motor Cycles*. Hamlyn Publishing Group Limited. (1973)

www.cardosystems.com/blog/the-history-of-the-motorcycle/

Cato, Jeremy "Harley-Davidson at 100". *The Vancouver Sun*. Vancouver, B.C. p. E.1.Fro. (8 August 2003)

Chelsea Area Historical Society. (2017)

https://chelseaupdate.com/1911-flanders-motorcycle-presented-chelsea-area-historical-society/ (August 2017)

Cycle and Automobile Trade Journal. (April 1909)

Cycle and Automobile Trade Journal. (March 1910)

Cycle and Automobile Trade Journal. (September 1910)

Detroit Public Library. Resource ID: bh023477 Flanders Factory. Digital Labs / Special Collections.

E-M-F Preliminary Catalog - The E-M-F Corporation. (1912)

Finney, E.J. Walter E. Flanders: His Role in the Mass Production of Automobiles, Published by Author 2000. (1992)

Flanders Mfg. Company. Flanders "4" Motorcycle, Advance Catalog. Pontiac, Michigan (1910)

Flanders, Paul. Flanders Company, Inc. (E-mail with author January 04, 2023)

Garson, Paul. Investing in Precious Metal. Hershey Vintage Motorcycle Auction. (2010)

https://getagripandmore.com/products/antique-mail-ad-ephemera-1913-flanders-motorcycles-harris-brothers-envelope

Hendry, Maurice M. Studebaker: One can do a lot of remembering in South Bend. New Albany: Automobile Quarterly. (1972)

Kimes, Beverly R. The Standard Catalog of American Cars 1805 - 1945 Kraus Publications. ISBN: 0-87341-428-4. (1996)

Leonard, Kevin. "Speedway's wooden track hosted auto races in the 20's" The Laurel Leader. (2012)

Library of Congress, US. Public Domain images.

Litwin, Matt. An Infant Titan Snuffed Out By Mistrust. Hemmings. (2018)

Longstreet, Stephen A. "A Century on Wheels: The Story of Studebaker" Henry Holt Company. (1952)

Marrin, Doug. Then & Now: Chelsea Clocktower's Fascinating History. The Sun Times News, Chelsea, Michigan (2017)

Mecum Auctions. Photo Credits (including cover photo). - as approved. (2022)

Mecum Auctions. 1914 Flanders Twin. Lot S111, Las Vegas. (2017)

Motor Age. (December 1908)

Motor Age. (January 1911)

Motorcycle Illustrated, various volumes. (1910)

Rafferty, Tod. "The Complete Illustrated Encyclopedia of American Motorcycles" Courage Books. (1999)

Reynolds, Cynthia Furlong. "Rip-Roaring Return. A 1911 Motorcycle Comes Home" Ann Arbor Observer. (2016)

Rubenstein, James M. "The Changing United States Auto Industry: A Geographical Analysis" Routledge. (1992)

https://searshomes.org/index.php/2011/02/10/those-darn-house-wreckers-harris-brothers/

Shavalia, Candice, Chelsea Area Historical Society photograph. (2017)

www.sfomuseum.org/exhibitions/early-american-motorcycles

http://www.speedwayandroadracehistory.com/baltimore-washington-speedway.html

St. Francis Motorcycle Museum. (2017)

Staff Writers, The Automobile. "E-M-F buys DeLuxe plant - Will build small car". (1909)

The Chelsea Standard newspaper, Chelsea Michigan. (January 12, 1911; February 16, 1911; September 5, 1912; July 10, 1913; July 24, 1913; July 31, 1913; October 3, 1913; October 16, 1913; October 8, 1914; August 26, 1915; June 21, 1923)

https://www.thehenryford.org/collections-and-research/digital-collections/archival-collections/403378/

Walker, Mick. *Motorcycle: Evolution, Design, Passion*. JHU Press. ISBN 978-0-8018-8530-3. (2006)

Wright, Stephen. "The American Motorcycle, 1869–1914" Megden Publishing Co. (2001)

Yanik, Anthony J. "The E-M-F Company: The Story of Automotive Pioneers Barney Everitt, William Metzger, and Walter Flanders" (SAE) ISBN: 0-7680-0716-X. (2001)

Youngblood, Ed. American Motorcyclist. Vol. 55, no. 6. American Motorcyclist Assoc. (June 2001)

Made in the USA
Middletown, DE
13 October 2023